Fighting Back the Sea

SCHOOL PUBLISHERS

Visit *The Learning Site!* www.harcourtschool.com

The Netherlands

The Dutch live with water. Their country, the Netherlands, is beside the North Sea. The location has three big rivers.

The land is flat and low. Half of it is lower than the sea. The country is also small. There is not much room.

The Dutch are the people of the Netherlands. Some call the place Holland.

In 1953, there was a lot of rain. Water covered the land.

Sometimes, the Dutch push back the water. They use the new land. They build homes and farms on it.

Sometimes, the sea fights back. Waves pound the shore. The rivers fill with rain. They try to spill over their banks. Then, the Dutch must fight the water.

The Netherlands Long Ago

The fight against the sea began long ago. The early Dutch had few resources.

The Dutch pushed back the water by hand. They piled earth to make walls, called dikes. The dikes held back water, and the ground dried out.

Two times, the sea pushed back hard!
In 1421 and 1570, the dikes broke. Water
covered the land.

The Dutch didn't give up. They made
their dikes higher. They used their windmills
to pump water from the fields.

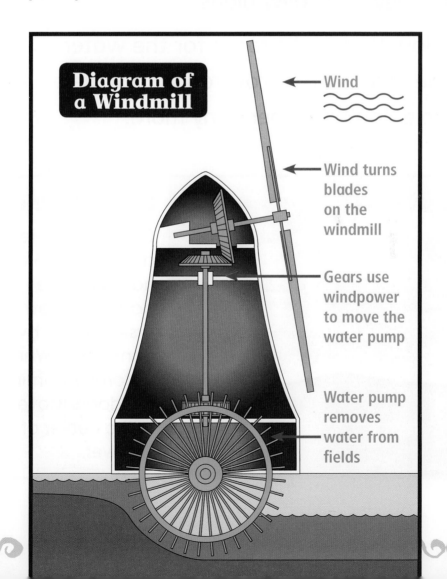

Diagram of a Windmill

Wind

Wind turns blades on the windmill

Gears use windpower to move the water pump

Water pump removes water from fields

The Netherlands Today

The Dutch still fight the sea. But things are different now. There is a new danger.

The sea is changing. It is getting warmer. The water is rising. The Dutch have changed, too. They have new plans.

Now they make room for the water. Some land is often covered with water. People don't live there any more.

Storm Barrier

One gate protects a port. The gate has two huge doors. The new system is run by computers. When a storm is coming, the doors move to the center of the river.

The Dutch have new, stronger dikes.
They are big gates. Most of the time, the
gates stay open. When the weather is bad,
they close. Then, the sea is pushed back.

The Dutch know how to live with water!

This is a new kind of dike.
It is a high gate, not a wall.

Think and Respond

1. What sea is beside the Netherlands?

2. How does the sea fight the Dutch?

3. How do the Dutch use windmills?

4. What would happen to the Netherlands without dikes?

 # Activity

Imagine that your family is visiting the Netherlands. Write a letter to a friend back home. Tell about your trip.